Impact of Education and Farm Size on ASF Knowledge Among Pig Farmers in Ghana. A Study

Mubarik Mahamud

© GRIN Publishing GmbH
Trappentreustraße 1
80339 München

Print and binding: Books on Demand GmbH, Norderstedt, Germany
Printed on acid-free paper from responsible sources.

The present work has been carefully prepared. Nevertheless, authors and publishers do not incur liability for the correctness of information, notes, links and advice as well as any printing errors.

GRIN web shop: https://www.grin.com/document/1436726

ANIMAL HEALTH AND PRODUCTION COLLEGE, PONG-TAMALE

ASSESSING THE KNOWLEDGE LEVE OF PIG FARMERS ON AFRICAN SWINE
FEVER IN JIRAPA MUNICIPALITY

MAHAMUD MUBARIK

A PROJECT WORK SUBMITTED TO ANIMAL HEALTH AND PRODUCTION
COLLEGE IN PARTIAL FULFILMENT OF THE REQUIREMENTS FOR AWARD OF
DIPLOMA IN ANIMAL HEALTH

MARCH 2023

ABSTRACT

ASFV causes a highly virulent disease in domestic pigs with a significant economic impact on the pig industry due to a fatality rate of up to 100% and the culling measures and movement bans necessary to control the virus. Containing the ASF remains a challenge, particularly due to the absence of vaccines or treatment, the stability of the virus in the environment, potential illegal movements of live pigs and pig products, the relatively large number of low-biosecurity farms, the commonplace use of swill feeding, and frequent interactions between domestic and wild swine. Hence, management and control efforts are mostly encouraged. The aim of this study was to determine the knowledge of ASF among pig farmers in the Jirapa municipality of the Upper East region of Ghana. The study adopted the. The cross-sectional design was used and a multi-stage sampling technique adopted to recruit 106 respondents. The minimum, maximum, mean, standard deviation and variance of the ages of the respondents were respectively 20 years, 51 years, 29.03 ($\pm$S.E.=0.806), 8.296 and 68.828. The minimum, maximum, mean, standard deviation and variance of the farm sizes from the study are 5, 123, 41.68 ($\pm$S.E.=2.669), 27.477 and 755.001 respectively. Majority (61.6%) of the respondents scored 0-2, indicating low knowledge, while 27.2% had scores of 5–9 signifying moderate knowledge and 11.1% scored 10–14, indicating high knowledge. The findings of the study further revealed that the level of education had a relationship with the gender, housing, her size and gender of the responding farmers. Farmers with moderate knowledge level were characteristically noted to be influenced by herd size ($P = 0.005<0.05$), housing ($P = 0.003<0.05$) and occupation of the farmers ($P = 0.000<0.05$). Farmers with poor knowledge level were shown to be influenced by the herd size ($P = 0.003<0.05$) of the farmers. In conclusion, most of them had merely heard of ASF but had no idea about what it entails, hence a relatively low knowledge level among the pig farmers about the infection among their flock.

DEDICATION

I dedicate the completion of this project work to my parents for their undying love and trust

in me throughout the years.

ACKNOWLEDGEMENT

All praises are due to God Almighty for granting me the life to accomplish this milestone feat. I also want to greatly appreciate my supervisor for his unnatural patience and guidance in the execution of this work. I further want to express my gratitude to the teaching and non-teaching staff of the Animal Health and Production College (Pong-Tamale) for the immeasurable roles they played in the completion of this work. I also want to thank the enumerator who assisted me in the collection of data for the study. Finally, I wish to express my heartfelt thanks to the ladies and gentlemen that assisted in the study

TABLE OF CONTENTS

List of Tables

CHAPTER ONE

1.0 INTRODUCTION

1.1 BACKGROUND TO THE STUDY

The circulation of the African swine fever virus (ASFV), since the first outbreak in 2007, has caused considerable concern over the potential for ASFV to spread into West Africa (Aliro et al., 2022). ASFV causes a highly virulent disease in domestic pigs with a significant economic impact on the pig industry due to a fatality rate of up to 100% and the culling measures and movement bans necessary to control the virus (Ackerman, 2022; Guinat et al., 2016). Containing the ASF remains a challenge, particularly due to the absence of vaccines or treatment, the stability of the virus in the environment, potential illegal movements of live pigs and pig products, the relatively large number of low-biosecurity farms, the commonplace use of swill feeding, and frequent interactions between domestic and wild swine (Ackerman, 2022; FAO, 2017). In ASF-free countries, early detection of ASFV introduction is, therefore, crucial to protect pig health, maintain access to global trade in pigs and pig products, and thereby limit the economic impact of an ASF epidemic.

In Ghana, ASF could potentially be introduced through the legal and illegal movement of contaminated pig products which may then also subsequently be used for swill feeding. Around 23% of pig farmers interviewed in Ghana have reported using swill to feed their pigs although this practice was banned in 2011 (Animal et al., 2018). In a recent study, this risk of introduction was considered to be one of the highest in the West African sub-region, due to a large number of ports, airports and travellers coming from affected areas (FAO, 2017). Therefore, the possibility of an ASF epidemic represents a significant economic risk, and preparedness for an ASFV incursion is of high importance to the Ghanaian pig industry. In Ghana, the current surveillance system for ASFV relies entirely on the immediate and

1

mandatory notification of any clinically suspect animals by pig farmers to their veterinary officers (Animal et al., 2018; FAO, 2017), who will then initiate an investigation into the menace. Blood and tissue samples are collected for laboratory analysis, and diagnostic results should be available within 24–48 hours. While under investigation, suspected farms are placed under movement restrictions to reduce the risk of ASFV spread via infected pigs, contaminated vehicles or equipment. In the case of ASFV introduction into the country, the compliance of all farmers with this policy should be able to limit the spread of the virus and thereby mitigate the early spread of the epidemic before control strategies can be implemented. However, a number of factors may delay reporting by farmers. Firstly, the symptoms of ASF are not specific, particularly in the early clinical stage of the disease. Fever, lethargy and loss of appetite are generally observed among infected domestic pigs, in association with sudden deaths within 2–3 weeks (Angel & Chueca, 2022; Lee et al., 2021; Penrith et al., 2022). Other symptoms can include vomiting, diarrhoea and haemorrhages. This is similar to other pig diseases, particularly classical swine fever (CSF), or porcine reproductive and respiratory syndrome (PRRS) which is endemic in Ghana (Animal et al., 2018). The disease can also express itself in some domestic pigs in more unusual clinical manifestations, such as with neurological symptoms, and in some cases may even be asymptomatic (Lee et al., 2021; Penrith et al., 2022). Previous surveys have concluded that farmers' awareness is biased towards diseases which occur at higher prevalence (Angel & Chueca, 2022), such as PRRS. Since an ASF outbreak has never been confirmed in Ghana, it is unlikely that farmers will have much awareness of the disease, if any. This view is supported by the fact that the Agriculture and Horticulture Development Board (AHDB Pork) recently had to cancel seminars on ASF and porcine epidemic diarrhoea virus due to a lack of interest from the pig industry (Angel & Chueca, 2022; Guinat et al., 2016). Farmers' awareness of specific clinical signs is essential for the early detection of any disease by state authorities (FAO, 2017). In addition, even if farmers are aware of a particular disease, whether and when they decide to report suspected cases will depend on a variety of factors. It is against this backdrop that this current study will be conducted in the Jirapa Municipality to measure pig farmers' level of knowledge of the African swine fever in pigs.

1.2   PROBLEM STATEMENT

ASFV is highly lethal, and an economically important factor in the pig production industry, since the virus can affect as high as 100% mortality in infected animals (Ackerman, 2022; FAO, 2017). Thus, the disease is a global canker. Its impact is especially severe in most developing countries where surveillance is poor (Ackerman, 2022; FAO, 2017). A previous study suggested that the main reasons for pig farmers not to immediately report suspected ASF cases included a lack of knowledge of reporting procedures, the perceived impact a notification could have on their reputation, the expectation that laboratory confirmation would take a long time, and the belief they could handle the outbreak themselves without the involvement of veterinary services (Ackerman, 2022; FAO, 2017; Guinat et al., 2016). A search of literature revealed extremely scanty studies to determine the knowledge and practices of pig farmers in most parts of Ghana regarding African Swine Fever. This observation is more pronounced in the Upper regions of Ghana, where pig production is rife but carried out in ignorance. It is against this observation that the current study will be conducted.

1.3   JUSTIFICATION OF THE STUDY

The ASF situation in Ghana appears to be fairly stable, with Ghana reporting outbreaks for the first time in 1999 (A. Rozstalnyy, B. Plavšić, 2019). Since 2009, the virus has been penetrating pig stocks year after year in the country. The last reported ASF outbreak was in Cape Coast, in the Central Region, in July 2018 (A. Rozstalnyy, B. Plavšić, 2019; Archibald, 2018). 898 pigs were culled in that outbreak (Archibald, 2018). The Jirapa municipality is the closest to the Burkina Faso/Ghana border, where transactions frequently occur and the risk of cross-border spread of diseases is extremely high. However, the prevalence of ASF is relatively extensive in Burkina Faso, according to Penrith et al. (2022). Sidi et al. (2022) report that there is a high seroprevalence of African swine fever in Burkina Faso with variations depending on the region and the breeding system. Thus, given the high transaction of business between the inhabitants of Southern Burkina Faso and the northernmost settlements of the upper regions of Ghana, it is utterly necessary to stress pig farmers' ability to identify pertinent information about ASF for potential outbreak detection and management.

1.4   RESEARCH OBJECTIVES

1.4.1   MAIN OBJECTIVE

To determine the knowledge level of pig farmers on African swine fever in Jirapa municipality.

1.4.2   SPECIFIC OBJECTIVE

I.   To examine the awareness of pig farmers of the African swine fever

II.   To determine the factors influencing pig farmers' level of awareness of African swine fever

1.5   RESEARCH QUESTIONS

1.5.1   Main research question

What is the knowledge level of pig farmers on African swine fever in Jirapa municipality?

1.5.2   Specific research questions

I.   What is the awareness of pig farmers of African swine fever?

II.   What factors influence pig farmers' level of awareness of African swine fever?

1.6   Significance of the Study

The study will be instrumental to understanding the knowledge-induced factors responsible for the outbreak and reoccurrence of the African Swine Influenza. It will present recommendations that could be useful in the improvement of existing policy interventions and the formulation of new policies against the prevalence of the disease. Thus, this study will furnish the government and other stakeholders with relevant information to assess the viability of interventions against the disease. Finally, this study will bridge the existing gap in research in this field of study, providing baseline data for future research.

1.7   Scope of the Study

The current study will focus on understanding the degree of knowledge of pig farmers of African Swine fever of pigs. The study will be carried out in the Jirapa municipality of the Upper East region of Ghana. This study has two specific focal areas. The first one is to measure the level of knowledge farmers of pigs have relating to African swine fever. The second focal point is to determine whether or not certain variables affect the perception f farmers of the disease.

1.8   Limitations and Delimitations of the Study

The researchers had limited time to conduct the survey and climax their study, so they needed to adopt the cross-sectional research design, which allow them to conduct the study within a very short period. Conversely, the results of the findings were not subjected to time and the impact the length of time will have on the knowledge and practices of the respondents about the disease. Therefore, the design will be a limitation of the study.

1.9   Organisation of the Thesis

The flow chart below illustrates the arrangement of the segments of the final thesis.

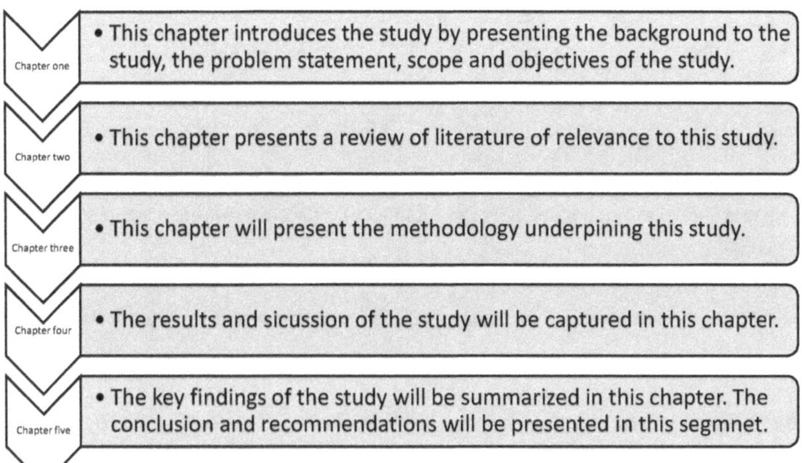

Figure 1 Organization of the thesis

CHAPTER TWO

2.0 LITERATURE REVIEW

2.1 INTRODUCTION

The literature review synthesises research findings of authorities in this area of study, and concepts that are relevant to fully explore the extent of knowledge available on this subject matter. This chapter consists of the conceptual literature review and the empirical literature review. The conceptual literature review presents key concepts that are relevant to understanding this problem better. The empirical literature review dissects the findings of other researchers on the level of knowledge of pig farmers of African swine fever.

2.2 CONCEPTUAL FRAMEWORK

2.2.1 History of African Swine Fever in West-Africa

The first official report to the OIE of ASF in West Africa was from Senegal in 1978, but a 1959 virus isolated from Dakar indicates that the virus was introduced at least two decades before (Penrith et al., 2022). The disease in West Africa appeared to remain restricted to southern Senegal and its neighbours (Guinea Bissau, Gambia and Cape Verde) until 1996 when Côte d'Ivoire experienced its first outbreak, and this was followed by an epidemic that involved most of the region's countries with any significant pig production (Benin, Nigeria, Togo, Ghana and Burkina Faso). The disease has since become endemic in most of these nations except Côte d'Ivoire, which achieved a prolonged eradication within a year until a new incursion in 2014 (Markosyan et al., 2019). Niger and Mali reported their first outbreaks in 2009 and 2016 respectively. No sylvatic cycle involving wild suids and/or Ornithodoros ticks

7

in maintaining the virus has been demonstrated. Only the genotype I am circulating, suggests the introduction rather than the evolution of the virus in the region (Chenais et al., 2019).

2.2.2 Epidemiology

The host range of ASFV is very narrow with suids as sole vertebrate hosts, and soft ticks of the genus Ornithodoros as competent arthropod vectors (Yoo et al., 2020). ASF has no zoonotic potential and there are no indications that this might change (Dixon et al., 2020). Among the reasons for the assumption that the evolution of a zoonotic potential is unlikely is the accurate proofreading of the DNA polymerase and virus-encoded base excision DNA repair system that leads to low mutation rates and the lack of possible recombination partners (FAO, 2017). The disease has its roots in sub-Saharan Africa where it is transmitted in an ancient sylvatic cycle among warthogs and Ornithodoros soft ticks, making ASFV the only DNA arthropod-borne (ARBO) virus. This cycle is not accompanied by overt disease or mortality in warthogs and would probably go unnoticed. Other African wild suids also show resistance (Blome et al., 2020). However, any introduction of the disease into the domestic pig sector via ticks or fomites leads to the above-described severe multi-systemic disease with high lethality. Ornithodoros soft ticks of another species complex (O. erraticus) were involved in outbreak scenarios on the Iberian Peninsula (Aliro et al., 2022; Animal et al., 2018). For the current global outbreak scenario, tick involvement does not seem to play any role. However, with the involvement of new countries and the emergence of tick species in new habitats, this might change. Once introduced into the domestic pig population, the virus does not need its arthropod vector for transmission. ASFV can be transmitted by direct contact between infected and susceptible animals and by indirect contact with contaminated objects or feed (Chenais et al., 2019). Contaminated pork (swill feeding) and also blood products used as protein sources can play an important role (Wen et al., 2019). Moreover, fomites such as clothing, trucks and veterinary equipment may act as sources of infection. In the wild boar habitat, carcasses are crucial in maintaining infection cycles (Busch et al., 2021; Sauter-louis et al., 2021). Moreover, persistently infected carrier animals have been discussed as an important factor for viral maintenance, especially in an endemic situation. The role of such persistently infected animals in the long-term transmission is still controversially discussed (Sauter-louis et al., 2021). Some of the controversies around the "persistence" of ASFV are probably rather a matter of definition. Beyond doubt, viruses and especially viral genomes can be detected in surviving animals for a rather long time (Lee et al., 2021). In the absence of truly neutralizing antibodies, the virus can still be isolated from survivors for roughly 60–70 days. Viral genome can be detected even longer (~100 days). However, there is no evidence for the major role of such carriers from field experience and long-term studies (Stahl et al., 2019). The latter showed a) no transmission to sentinels and b) no virus in survivors beyond 100 days (Busch et al., 2021; Lee et al., 2021). For surveillance actions, different categories of animals should be clearly defined: Animals that are positive only in pathogen detection methods should be regarded as animals in the early phase of infection (FAO, 2020; Viltrop et al., 2021). With high certainty, these animals can transmit the virus. Animals that show both viruses and antibodies have a certain probability to shed and transmit viruses. From seven to ten days post-infection to almost 100 days post-infection, animals can show such behaviour. As antibodies do not have a clear predictive value for disease outcome, these animals can still die from ASF, but they have at least survived for one week or more. The last category is animals with only antibodies. These animals are the true long-term survivors and should not be called persistently infected, as no virus was detected in relevant samples. These animals are probably protected from reinfection and thus very safe. It cannot be excluded that a couple of genome copies are still present somewhere in the lymphatic tissues. However, given the rather high dose usually needed for oral infection, i.e. 10,000 heam adsorbing units (Ivars, 2017), the impact should be low.

However, both the duration of immunity and possible reactivation scenarios need further investigation.

2.2.3 Transmission

The ASF virus persists in distinct cycles – traditionally, the sylvatic cycle, the tick-pig cycle and the domestic cycle. More recently, a wild boar cycle has been described, which may sometimes be involved in the latter (Markosyan et al., 2019). The sylvatic cycle occurs only in parts of Africa and involves warthogs and ticks of the Ornithodoros moubata complex. The tick-pig cycle involves pigs and Ornithodoros spp. ticks, which have been described as infesting parts of Africa and the Iberian Peninsula (Ackerman, 2022). Transmission from the sylvatic cycle to the domestic cycle occurs via indirect transmission by ticks. This can happen where pigs and warthogs share common grounds, particularly when warthogs establish burrows on farms, or when ticks are brought back to villages through the carcasses of warthogs killed for food (Viltrop et al., 2021).

2.2.4 Clinical signs in susceptible hosts

The clinical signs of ASF are highly variable and depend on the virulence of the strain and the age and immune status of the animals (Chenais et al., 2019). Apart from acute diseases resembling haemorrhagic fever, chronic and subclinical courses can occur. The causative virus strains in Europe (apart from Sardinia) and Asia are of genotype II, highly related, and show high virulence for both domestic pigs and European wild boar under experimental conditions (Busch et al., 2021; FAO, 2020). Highly virulent strains cause acute to peracute disease with up to 100 % lethality within 7–10 days. The clinical signs are often non-specific and include high fever, anorexia, respiratory and gastrointestinal signs, cyanosis, ataxia, and peracute death. Pregnant sows can abort due to severe disease and high fever. Hemorrhagic symptoms have also been observed in a few cases. The experimental studies have been reviewed and summarized in two publications recently (Pikalo et al., 2019; Sanchez-Cordon et al., 2019).

Moderately virulent strains lead to an acute clinical picture with high fever, anorexia, fatigue and non-specific respiratory and gastro-intestinal symptoms. Pregnant animals can abort. The mortality rate, in this case, is 30–70 % (Lee et al., 2021; Viltrop et al., 2021). Low virulent strains show subclinical and chronic courses with unspecific symptoms and low mortality. Antibodies are formed after 7–10 days, but these are not predictive of disease out- come and are not able to completely neutralize the virus.

2.2.5   Control options

A binding legal framework exists for surveillance and control in most countries with substantial pig production. Integral parts of the control measures are timely and reliable diagnosis, stamping out of infected herds, the establishment of restriction zones, movement restrictions, and tracing of possible contacts (Muhangi et al., 2015). Prophylactic vaccination and other treatments are still not available but would be strictly prohibited in the EU and other countries. An example of the legal framework is the legislation at the EU level that is based on the experience gained during the outbreaks in the 1960ies to 1990ies (FAO, 2020). While it may not be applicable in resource-limited settings, the basic principles can be applied worldwide. For this reason, some general points are outlined here. At present, Council Directive 2002/60/EC lays down the minimum Community measures for the control of ASF (Sauter-louis et al., 2021). Chapters are dedicated to measures in case of suspicion, measures in case of a confirmed outbreak, measures in contact holdings and epidemiological inquiries, the establishment of protection and surveillance zones, cleaning and disinfection, measures that are taken to lift restrictions and for repopulation, and details special cases such as suspicions at slaughterhouses. Commission Decision 2003/422/EC details the control measures and provides guidelines and minimum requirements on diagnostic procedures, sampling methods and criteria for evaluation of the results, biosafety requirements, and principles and applications of laboratory tests. The legal framework at the EU level is under revision and will be replaced by a new animal health law in 2021. However, the basic principles will stay in place (Beltrán-Alcrudo, D.;Arias, M.; Gallardo, C.;Kramer, S.;Penrith, 2017; Guinat et al., 2016).

Implementation of control measures depends on the resources of the veterinary service and compliance with these measures. A critical point for compliance with measures is a timely and appropriate compensation scheme. As can be seen in Africa and Asia, the lack of compensation leads to the selling of sick pigs to slaughterhouses and markets resulting in large outbreak scenarios (Dixon et al., 2020). For the wild boar cycle, local situations must be taken into consideration when designing control measures. Currently, the example of the Czech Republic is often used, as this is the only case where the disease was completely eradicated from a country's wild boar population (so far). Based on these successful measures (EFSA, 2018), a control strategy could involve the following: Zonation determining an infected zone and surrounding buffer and control zones. These zones should be established as soon as possible and should be sufficiently big. Carcasses should be searched and disposed of. Incentives for carcasses found can increase compliance and success. In the infected zone, strict hunting rest should be applied. Trapping can be allowed if established and appropriate. Fencing has proven successful and can be applied and adapted to the local situation. Fences are not pig proof but can slow down the spread to a minimum. Where possible, entry bans can be implemented. Reduction of wild boar in the surrounding area can be applied but must be decided individually for each case.

2.2.6 African swine fever vaccines

To date, safe and efficacious vaccines against ASF are still lacking and production of most vaccine candidates is massively hampered by the simultaneous lack of a permanent cell line that is sufficiently susceptible to ASFV and does not force further genetic adaptations within the ASFV genome. However, vaccine design seems still feasible since animals recovering from acute ASF are protected against challenging infection with closely related strains (FAO, 2017; OIE, 2020). Problems in developing a vaccine are partly because antibodies produced by the affected animal cannot render the virus harmless (no complete neutralization). Furthermore, many components that would be important for the establishment of a favourable immune response are not yet fully characterized in localization and function, so it is challenging to produce vaccines based on individual components or antigens expressed by vectors. Inactivated virus preparations have shown no substantial protective effect, although antibodies were formed against the virus (Animal et al., 2018). Finally, it should also be mentioned that the virus can establish complex immune modulation and efficient immune evasion (Reis et al., 2017; Dixon et al., 2013), i.e., it brings along numerous factors that influence the host's immune system in such a way that virus replication can take place efficiently. With the pandemic spread of ASF, research towards vaccine development has been intensified and some very promising results have been obtained. ASFV vaccines may be closer than they appear (Bosch-Camós et al., 2020). As several recent reviews summarize the approaches to creating safe and efficient vaccines against ASF (SUIDE, 2022), the section below is mainly focused on general features, recent publications and developments, and also some rather neglected issues.

2.3 EMPIRICAL REVIEW

In a study conducted on English pig farmers, Guinat et al. (2016) discovered that pig farmers having poor knowledge about ASF clinical signs and limited concern about ASF compared with other pig diseases are less likely to consider the possibility of an outbreak of ASF on their farm. In addition, pig farmers lacking awareness of outbreaks in other countries, having a perception of the negative impact on them resulting from false positive reporting and the perceived complexity of reporting procedures are less likely to report an ASF suspicion. Guinat et al. (2016) proposed the need for educational campaigns targeted at pig farmers to focus on in an attempt to increase the likelihood of a rapid response in the event of an ASF outbreak.

Aliro et al. (2022) found in their study that pig farmers mostly had positive perceptions of ASF biosecurity, describing measures as effective. Participants further possessed knowledge of ASF and its transmission, some of which were in line with known scientific knowledge and some not. Nevertheless, the farmers were hindered from preventing and controlling ASF due to biosecurity costs and a need to prioritise family livelihood over disease transmission risks, incompatibility of current biosecurity practises with local culture, traditions and social contexts and finally lack access to veterinarians or, occasionally, low-quality veterinary services (Aliro et al., 2022). The paper recommends that participatory processes be applied in designing biosecurity measures to ensure better adaptation to local cultural and social contexts

3.0 METHODOLOGY

3.1.1 Introduction

This chapter forms the framework within which the study will be conducted. This includes a description of the research design, sample collection technique, data collection and analysis. This chapter will usher the researcher into the subsequent chapters of this work.

3.1.2 Study area

This current study will be conducted in the Jirapa municipality between the period of January 2023 and March 2023. The Municipality is located in the northwestern corner of the Upper West Region of Ghana and is one of eight Municipalities in the region. It lies approximately between latitudes 10.25o and 11.00o North and longitudes 20.25o and 20.40o West with a territorial size of 1,188.6 square kilometres representing 6.4 per cent of the regional landmass. (GSS, 2014). Jirapa Municipality is bordered to the south by the Nadowli-Kaleo Municipality, to the north by the Lambussie-Karni Municipality, to the West by Lawra Municipality and to the east by the Sissala West Municipality. The Municipality capital, Jirapa, is 62 km away from Wa, the regional capital. Its location presents a special development advantage for the Municipality (GSS, 2014). The population of Jirapa Municipality, according to the 2010 Population and Housing Census, is 88,402 representing 12.6 per cent of the region's total population. Males constitute 47.0 per cent and females represent 53.0 per cent. About 85.6 per cent of the population lives in rural localities (GSS, 2014).

3.2 STUDY POPULATION

The target population is a subset of the population, and it is well-defined. Accordingly, the target population of this study would consist of pig farmers in the Jirapa municipality of the Upper East region of Ghana.

3.3 STUDY DESIGN

The research design adopted for this study is the survey design. The guiding principles of such designs is that it is suitable for easy data accessibility, especially in cases where the researcher is financially limited (Ali, 2018; Saxena et al., 2013).

3.4 SAMPLE SELECTION TECHNIQUE AND SAMPLE SIZE

The sample size will be selected through a two-step sampling technique. The first stage is the selection of 2 communities (Jirapa and Tizza) in the Jirapa Municipality. Subsequently, the stratified sampling technique will be used to select 53 respondents from each community. Thus, 106 pig farmers will be recruited for this study.

3.4.1 Instrumentation of questionnaire

The principal tool for primary data collection for this study is the questionnaire. The questionnaire will be designed into three (3) distinct sections (Sections A, B and C) in accordance with the objectives of this study. Section A will elicit information on the sociodemographic characteristics of the responding farmers. Section B will present questions on the husbandry practices of the farmers and sections C will assess the level of knowledge of ASF among the farmers.

3.4.2 Survey

The survey will be commenced upon receiving an introductory letter from the Animal Health and Production College, permitting me to carry out the study. The introductory letter will be subsequently presented to the target respondents and their autonomous permission and willingness to assist with this study will be sought. Respondents who are permitted to partake in this study will be administered the questionnaire. However, respondents were fully made to understand that their responses would not be published with their identities and that they reserved all the rights to excuse themselves from the study midway. The questionnaire was interpreted in Dagarti (the predominantly spoken local language in the area) for the understanding of the respondents who could not understand the English language (the default language in which the questionnaire will be drafted).

3.5   TESTING FOR RELIABILITY AND VALIDITY OF THE QUESTIONNAIRE

The reliability of the structured questionnaires will be checked to ensure that it is structured appropriately and capable of obtaining data of relevance to the study. Therefore, 5 respondents will be administered the questionnaires in a pre-test and an equal number of different respondents from the sample size in a post-test. To ensure that the results of this study are reliable and accurate, the survey questionnaires were adjusted to obtain the most relevant data from the respondents. To accomplish this feat, an array of questions will be initially administered to a proportion of the population. The most relevant questions will be recruited to form the final questionnaires for data collection.

3.6   DATA ANALYSIS

Data collected during the survey will be input into Microsoft Excel and coded. Coded data will be analysed using the IBM Statistical Package for the Social Sciences (SPSS) version 26. Analysed data will be displayed on tables and charts and the frequencies, percentages, means, and standard deviations of the data interpreted.

4.0 RESULTS AND DISCUSSION

4.1 INTRODUCTION

This chapter presents the results obtained from the analysis of the survey data. It also presents

an objective discussion of the key discoveries in relation to the existing literature as captured

in chapter two of this report. This chapter is segregated into sections that mimic the specific

objectives of the study, to ease comprehension.

4.2 DESCRIPTIVE STATISTIC

Table 1 Descriptive Statistics

Descriptive Statistics	Minimum	Maximum	Mean	Std. Error	Std. Deviation	Variance
Age	20	51	29.0	0.8	8.3	68.9
Herd Size	**5**	**123**	**41.7**	**2.7**	**27.5**	**755.0**

The minimum, maximum, mean, standard deviation and variance of the ages of the respondents

as shown in Table 1 are respectively 20 years, 51 years, 29.0 (±S.E.=0.806), 8.3 and 68.9. The

minimum, maximum, mean, standard deviation and variance of the herd sizes from the study

are 5, 123, 41.7 (±S.E.=2.669), 27.5 and 755.0 respectively. The mean age of the farmers

suggests that pig farming is predominantly patronised by young people, indicating its

importance in employing the most vibrant age group in the study area.

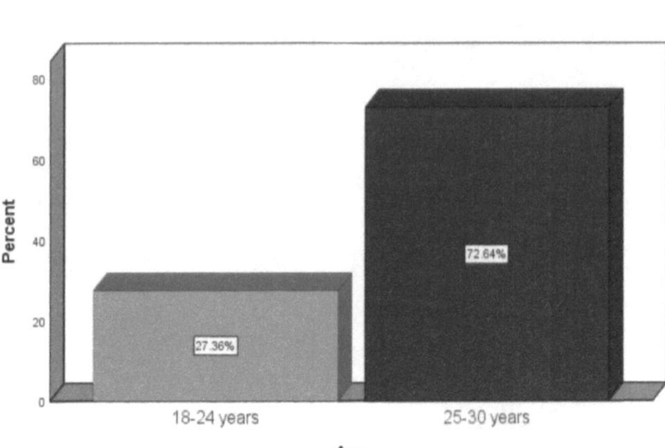

Figure 2 Age of Respondents

The distribution of the respondents by age is presented in Figure 2. The results show that

72.64% of the respondents were aged 25-30 years. 27.36% of the respondents were aged 18-

24 years.

Table 2 Gender and Occupation of Respondents

	Variable	Frequency	Per cent
	Female	29	27.4
Gender	Male	77	72.6
	Total	106	100
	Farmer	44	41.5
	Trader	7	6.6
Occupation	Government worker	21	19.8
	Unemployed	34	32.1
	Total	**106**	**100**

The results on the gender and occupation of the respondents reveal that the majority (72.6%)

of the respondents were males, while 27.4% of them were females. The majority of the

respondents engaged in farming (41.5%), 32.1% of them were unemployed, 19.8% were

government workers and traders constituted 6.6% of the respondents. The findings of the study

reveal that pig farming is a male-dominated activity, confirming reports by the Ghana

Statistical Service that agriculture is poorly patronised by females in the Jirapa municipality

(GSS, 2014).

Table 3 Farm Information

Variable		Frequency	Per cent
	Less than 10	10	9.4
	10-49	63	59.4
Herd size	50-100	30	28.3
	Over 100	3	2.8
	Total	**106**	**100**
	Indoor	38	35.8
Housing	Outdoor	68	64.2
	Total	**106**	**100**

In Table 3, most of the farmers owned between 10 and 49 pigs (59.4%, 63). The majority of

farmers adopted the outdoor system of housing their pigs (64.2%, 68). The findings of the study

corroborate the empirical report by the Ghana Statistical Service that the majority of pig

producers in the Jirapa municipality practice an extensive farming system (GSS, 2014).

Table 4 Knowledge of African Swine Fever

Variable		Frequency	Per cent
	Yes	28	26.4
Heard of ASF	No	78	73.6
	Total	106	100
	Yes	39	36.8
Know about the spread of ASF	No	67	63.2
	Total	106	100
	Yes	26	24.5
Know the clinical signs of ASF	No	80	75.5
	Total	106	100
	Yes	55	51.9
Infected can spread ASF	No	51	48.1
	Total	106	100
	Yes	30	28.3
Disease is hereditary	No	62	58.5
	I don't know	14	13.2
	Total	106	100
	Yes	47	44.3
Can healthy pigs be vaccinated	No	46	43.4
	I don't know	13	12.3
	Total	106	100

		Frequency	Per cent
	Yes	28	26.4
Is there a treatment for ASF	No	63	59.4
	I don't know	15	14.2
	Total	**106**	**100**

The findings also revealed that 73.6% of the respondents were unaware of ASF, and 36.8% of the study participants indicated ASF infection could be transmitted. It was also observed that 51.9% did not know whether ASF infection could be spread by an infected person (Table 4). It was discovered that 75.5% of the respondents were unaware of the clinical signs of ASF; 28.3% of them stated that the infection was hereditary. The study further revealed that 59.4% of the participants did not know if there was a known treatment for ASF. 44.3% of the respondents said unhealthy pigs could be vaccinated against ASF, and 59.4% of them disclosed that there was a treatment against ASF.

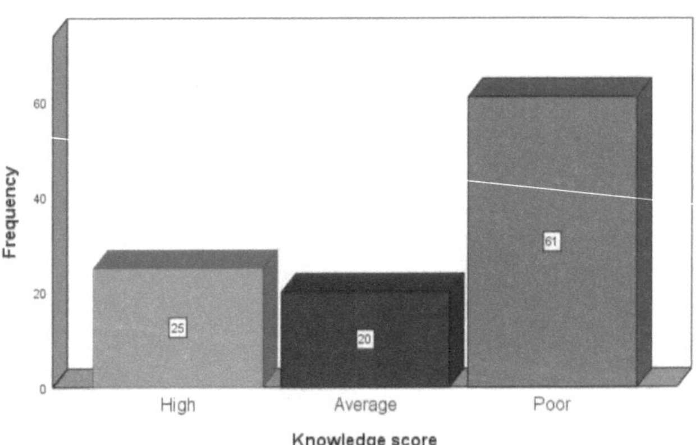

Figure 3 Knowledge Score

Table 5 Knowledge Score

Knowledge score	Frequency	Per cent
High	25	23.6
Average	20	18.9
Poor	61	57.5

| **Total** | 106 | 100 |

Infection. Respondents were rated according to their responses to the questions pertaining to knowledge (Table 5; Figure 3), which were scored as "1" for a correct response and "0" for a wrong response and subsequently summed up for each respondent (composite scoring). A majority (57.5%) of the respondents scored between 0 and 2, indicating low knowledge, while 18.9% had scores within the range of 5–9 signifying moderate knowledge whilst 23.6% scored 10–14, indicating high knowledge. The findings of this study corroborate that of Guinat et al. (2016) who discovered that pig farmers had poor knowledge about ASF clinical signs and the mechanism of transmission of the disease on their farms. However, the findings of this study further confirm that of Viltrop et al. (2021), where it was divulged that pig farmers generally lack sufficient knowledge of ASF across the globe.

4.4 CORRELATION TEST OF ASSOCIATION

A correlation test is used to evaluate the association between the knowledge score (high, moderate or poor) and sociodemographic and farm parameters to establish the relationship between these variables. There are different types of correlation tests such as Pearson correlation coefficient, Spearman's rank correlation coefficient, Kendall's rank correlation coefficient and others. In this study, the Pearson correlation was used to assess the relationship between these variables. The Pearson correlation is appropriate for this study because the test variables are quantitative and/or nominal. The results obtained from the statistical test is presented in Table 6 below. The results indicate that there was no relationship between high ASF knowledge level and gender ($P = 0.112 < 0.05$), herd size ($P = 0.194 < 0.05$), housing ($P = 0.11 < 0.05$) and occupation of the farmers ($P = 0.104 < 0.05$) were statistically insignificant. Thus, there was no effect of their gender, herd size, housing and occupation of the farmers on whether or not their level of education was high.

Table 6 Correlation Test of Association

Correlations		Gender	Herd size	Housing	Occupation
high	Pearson correlation	1	0.121	0.152	-0.151
	Sig. (2-tailed)	0.112	0.194	0.110	0.104
	N	106	106	106	97
Moderate	Pearson correlation	0.121	1	0.282	-0.328
	Sig. (2-tailed)	0.194	0.005	0.003	0.000
	N	98	106	106	93
Poor	Pearson correlation	0.152	0.282	1	-0.534
	Sig. (2-tailed)	0.110	0.003	-	0.000
	N	**96**	**96**	**97**	**95**

The results indicate that there was no relationship between moderate ASF knowledge level and gender (P = 0.194<0.05). However, a test of the relationship between moderate knowledge level and herd size (P = 0.005<0.05), housing (P = 0.003<0.05) and occupation of the farmers (P = 0.000<0.05) yielded a statistically significant association. Thus, the herd size, housing and occupation of the responding farmers strongly associates with moderate level of knowledge of ASF. Finally, the gender (P = 0.11<0.05), herd size (P = 0.003<0.05) and occupation (P = 0.194<0.05) of the farmers was discovered to exhibit poor knowledge of ASF at a level of significance of 0.05.

5.0 SUMMARY, CONCLUSION AND RECOMMENDATIONS

5.1 INTRODUCTION

This chapter presents the summary, conclusion and recommendations of this study. The summary recapitulates the key findings of this study. The conclusion is an overall statement on the findings of this study. The recommendations are the suggestions derived from the findings of this study

5.2 SUMMARY

The purpose of this study was to assess the level of knowledge of African Swine Fever among pig farmers in the Jirapa municipality of the Upper East region of Ghana. The study conceived the following objectives. The first objective was to determine the level of awareness of ASF in pigs. The second research objective determined the factors affecting the level of knowledge of ASF among pig farmers in the study area. Data were collected and analysed by administering a questionnaire to 106 pig producers in the municipality.

The findings of the study divulged that the majority of the respondents were generally unaware of African Swine Fever. The propensity and mode of transmission of the disease were largely unknown to the pig farmers. The proportion of the respondents who did not know the mode and route of spread of ASF, could not decipher whether or not n infected pig could transmit the disease. The clinical manifestation of ASF was unknown to a majority of the respondents, indicating a very high chance of misdiagnosis and mistreatment without the aid of a veterinary professional. ASF is considered an inborn disease among 57.5% of the respondents, and knowledge of the availability of an approved treatment for ASF was lacking among 67% of the respondents. Overall, the Majority (61.6%) of the respondents scored between 0 and 2, indicating low knowledge, while 27.2% had scores within the range of 5–9 signifying moderate knowledge whilst 11.1% scored 10–14, indicating high knowledge. The findings of the study further revealed that the level of education had a relationship with the gender, housing, her size and gender of the responding farmers. In essence, the findings of the study disclosed that farmers with moderate knowledge level were characteristically noted to be influenced by herd size (P = 0.005<0.05), housing (P = 0.003<0.05) and occupation of the farmers (P = 0.000<0.05). Farmers with poor knowledge level were shown to be influenced by the herd size (P = 0.003<0.05) of the farmers.

5.3 CONCLUSION

Most of them had merely heard of ASF but had no idea about what it entails, hence a relatively low knowledge level among the pig farmers about the infection among their flock. Even though the majority of the respondents had some form of education, there was no significant association between their educational levels and level of knowledge of ASF in pigs. There was generally a low level of knowledge of the ASF among the study participants.

5.4 RECOMMENDATIONS

Based on the findings in the study, it is recommended that the Ghana Health Service and Ministry of Health carry out a free compulsory market-to-market vaccination program, while Municipal Assembly and District Health Directorate should organize and implement educational programmes from time to time to sensitize market women and the entire municipality about the HBV infection. Further studies are recommended to include other vulnerable workers and men to confirm the high HBV prevalence.

REFERENCES

A. Rozstalnyy, B. Plavšić, F. & O. (2019). *Strategic challenges to global control of African Swine Fever. 87th General Session, 26-31.05.2019. Paris. 33*(0), 26–31.

Ackerman, D. (2022). *2022 African Swine Fever Virus Research Review 2022 African Swine Fever Virus Research Review. 727494.*

Adam, A. M. (2020). Sample Size Determination in Survey Research. *Journal of Scientific Research and Reports, 26*(5), 90–97. https://doi.org/10.9734/jsrr/2020/v26i530263

Ali, M. (2018). *Introduction to study designs Today ' s presentation.* 10–35.

Aliro, T., Chenais, E., Odongo, W., Okello, D. M., Masembe, C., & Ståhl, K. (2022). *Prevention and Control of African Swine Fever in the Smallholder Pig Value Chain in Northern Uganda : Thematic Analysis of Stakeholders ' Perceptions. 8*(January). https://doi.org/10.3389/fvets.2021.707819

Angel, M., & Chueca, M. (2022). *Research objectives to fill knowledge gaps in African swine fever virus survival environment : 05 PM a Research objectives to fill knowledge gaps in African swine fever virus survival in the environment and carcasses , which could improve the control of A.* 1–19.

Animal, S., Veterin, M., Vej, F. B., Sciences, N., Technologies, I., Game, S., Real, C., Diseases, W., Sciences, B., Health, V. P., & Sciences, A. (2018). *1 . African swine fever (ASF), the pig health challenge of the century.* 11–24. https://doi.org/10.3920/978-90-8686-910-7

Archibald, D. (2018). *Ghana Ghana Confirms Outbreak of African Swine Fever. May*, 2017–2019.

Bartlett II, J. E., Kotrlik, J. W., & Higgins, C. C. (2001). Determining appropriate sample size in survey research. *Information Technology, Learning, and Performance Journal, 19*(1), 43–50.

Beltrán-Alcrudo, D.;Arias, M.; Gallardo, C.;Kramer, S.;Penrith, M. L. . (2017). African swine fever: detection and diagnosis. In *Food and Agriculture Organization of the United Nations, Rome* (Issue June).

Blome, S., Franzke, K., & Beer, M. (2020). African swine fever – A review of current knowledge. *Virus Research, 287*(August), 198099. https://doi.org/10.1016/j.virusres.2020.198099

Bosch-Camós, L., López, E., & Rodriguez, F. (2020). African swine fever vaccines: A promising work still in progress. *Porcine Health Management, 6*(1), 1–14. https://doi.org/10.1186/s40813-020-00154-2

Busch, F., Haumont, C., Penrith, M., & Laddomada, A. (2021). *Evidence-Based African Swine Fever Policies : Do We Address Virus and Host Adequately ? 8*(March), 1–12. https://doi.org/10.3389/fvets.2021.637487

Chenais, E., Depner, K., Guberti, V., Dietze, K., Viltrop, A., & Ståhl, K. (2019). Epidemiological considerations on African swine fever in Europe 2014-2018. *Porcine Health Management, 5*(1), 1–10. https://doi.org/10.1186/s40813-018-0109-2

FAO. (2017). *African swine fever (ASF) detection and diagnosis.*

FAO. (2020). African swine fever in wild boar ecology and biosecurity. In *African swine fever in wild boar ecology and biosecurity*. https://doi.org/10.4060/ca5987en

GSS. (2014). *Jirapa district*.

Guinat, C., Wall, B., Dixon, L., & Pfeiffer, D. U. (2016). English pig farmers' knowledge and behaviour towards African swine fever suspicion and reporting. *PLoS ONE, 11*(9), 1–13. https://doi.org/10.1371/journal.pone.0161431

Ivars, M. J. (2017). African swine fever (ASF). 生化学, *7*(3), 213–221.

Lee, H. S., Bui, V. N., Dao, D. T., Bui, N. A., Le, T. D., Kieu, M. A., Nguyen, Q. H., Tran, L. H., Roh, J., So, K., Hur, T., & Oh, S. (2021). *Pathogenicity of an African swine fever virus strain isolated in Vietnam and alternative diagnostic specimens for early detection of viral infection.* 1–11.

Markosyan, T., Sargsyan, K., Kharatyan, S., Elbakyan, H., Hakobyan, V., Simonyan, L., Voskanyan, H., Shirvanyan, A., Stepanyan, T., Khachatryan, M., Karapetyan, M., Avagyan, A., Mcvey, W. R., Weller, R., Keen, J., & Risatti, G. R. (2019). The epidemiological status of African swine fever in domestic swine herds in the Tavush Province region, Armenia. *Revue Scientifique et Technique (International Office of Epizootics), 38*(3), 751–760. https://doi.org/10.20506/rst.38.3.3024

Muhangi, D., Masembe, C., Emanuelson, U., Boqvist, S., Mayega, L., Ademun, R. O., Bishop, R. P., Ocaido, M., Berg, M., & Ståhl, K. (2015). A longitudinal survey of African swine fever in Uganda reveals high apparent disease incidence rates in domestic pigs, but absence of detectable persistent virus infections in blood and serum. *BMC Veterinary Research, 11*(1), 1–11. https://doi.org/10.1186/s12917-015-0426-5

OIE. (2020). *African swine fever: An unprecedented global threat A challenge to livelihoods, food security and biodiversity* (Issue October).

Penrith, M., Heerden, J. Van, Heath, L., Abworo, E. O., & Bastos, A. D. S. (2022). *Review of the Pig-Adapted African Swine Fever Viruses in and Outside Africa.*

Sauter-louis, C., Conraths, F. J., Probst, C., Blohm, U., Schulz, K., Sehl, J., Fischer, M., Forth, J. H., Zani, L., Depner, K., Mettenleiter, T. C., Beer, M., & Blome, S. (2021). *African Swine Fever in Wild Boar in Europe — A Review.*

Saxena, P., Prakash, A., Acharya, A. S., & Nigam, A. (2013). Selecting a study design for research. *Indian Journal of Medical Specialities, 4*(2). https://doi.org/10.7713/ijms.2013.0033

Sidi, M., Zerbo, H. L., Ouoba, B. L., Settypalli, T. B. K., Bazimo, G., Ouandaogo, H. S., Sie, B. N., Guy, I. S., Adama, D. D. tombo, Savadogo, J., Kabore-Ouedraogo, A., Kindo, M. G., Achenbach, J. E., Cattoli, G., & Lamien, C. E. (2022). Molecular characterization of African swine fever viruses from Burkina Faso, 2018. *BMC Veterinary Research, 18*(1), 1–11. https://doi.org/10.1186/s12917-022-03166-y

SUIDE. (2022). *https://www.woah.org/fileadmin/Home/eng/Health_standards/tahm/3.09.01_ASF.pdf 1.* 1–18.

Taherdoost, H. (2018). Sampling Methods in Research Methodology; How to Choose a Sampling Technique for Research. *SSRN Electronic Journal, September.*

https://doi.org/10.2139/ssrn.3205035

Viltrop, A., Boinas, F., Jori, F., & Kolbasov, D. (2021). Understanding and combatting African Swine Fever. In *Understanding and combatting African Swine Fever* (Issue March). https://doi.org/10.3920/978-90-8686-910-7

Yoo, D., Kim, H., & Lee, J. Y. (2020). *African swine fever : Etiology , epidemiological status in Korea , and perspective on control. 21*(2), 1–24.

Appendix 1 Questionnaire

ANIMAL HEALTH AND PRODUCTION COLLEGE

Introduction: I am a final year top-up student of the Animal Health and Production College (Pong-Tamale) and I am conducting a study on the ASSESSING THE KNOWLEDGE LEVEL OF PIG FARMERS ON AFRICAN SWINE FEVER. This forms part of requirements for the award of a Diploma certificate in Animal Health and Production. Your assistance, by responding to this questionnaire would be significant at assisting my project work. All your responses will be treated as confidential and will only be used for academic purposes.

Sociodemographic characteristics

1. Age   [.................................]
2. Gender
   Male [   ]   Female [   ]
3. Marital Status
   Single [   ]   Married [   ]   Divorced [   ]
4. What is your highest level of formal educational
   Uneducated [   ]   Basic [   ]   Secondary [   ]   Higher [   ]
5. What is your religious denomination?
   Christianity [   ]   Islam [   ]   ATR [   ]
6. Occupation
   Farmer [   ]   Trader [   ]   Government worker [   ]   Unemployed [   ]

Husbandry information

7. What scale is your production of small ruminant?
   Small-scale [   ]   Commercial [   ]
8. Heard size
   Less than 10 [   ] 10-49 [   ] 50-100 [   ] Over 100 [   ]

9. Housing system

Indoor [   ] Semi [   ] Outdoor [   ]

Knowledge of ASF

Tick against the row labelled "YES" or "NO" depending on your response to the statements in the left row below.

Statement	Yes	No
10. Heard of ASF		
11. Know about the spread of ASF		
12. Know the clinical signs of ASF		
13. Infected can spread ASF		
14. Disease is hereditary		
15. Can healthy pigs be vaccinated		
16. Is there a treatment for ASF		

Thank you